007

ガンツウ │ guntû

ガンツウ｜guntû

堀部安嗣
Yasushi Horibe

写真
鈴木研一
Photographs
Ken'ichi Suzuki

001 Chapter I 瀬戸内とガンツウ
Setouchi and guntû

044 Chapter II 瀬戸内の魅力を掘り起こすために
Uncovering the Allure of Setouchi

065 Chapter III ドローイング
Drawings

074 Chapter IV ガンツウの全貌
Overview of guntû

098 Chapter V 知られざる瀬戸内
Unknown Setouchi

瀬戸内の魅力を掘り起こすために

堀部安嗣

旅と乗り物の原風景
————

私は建築の設計をしていますが、最初に夢中になったのは乗り物を描くことでした。幼少期、港町横浜の鶴見に住んでいたので、鉄道や船には日常的に親しみを感じる環境にありました。育った祖母の家の窓からは、港を行き交う船や京浜間を走る鉄道が見えました。それらを見ながらいつも乗り物の絵を描き、そして乗り物に乗ることも好きになってゆきました。特に思い出深いのは、毎年夏休みの間、曾祖母が住んでいた高知へ行って1カ月ほど過ごしていたことです。新幹線ではなく当時「ブルートレイン」といわれていた寝台列車に乗って岡山の宇野まで行き、そこから宇高連絡船に乗って四国高松へ渡り、さらに気動車（ディーゼル車）に乗るという船と鉄道の長旅で

した。今はなき連絡船やブルートレインにはなんともいえない旅情がありました。また出発の翌朝は、連絡船の甲板にて、瀬戸内海の風に吹かれながら讃岐うどんが食べられるという特典もついていました。風景と空気がよければおいしさは倍増するのです。それは瀬戸内との初めての出会いでもあり、乗り物と風景と食が一体となった自分のなかでのひとつの原風景となっています。宇高連絡船では頻繁に行き交う船や無数の島々の眺めが印象的でしたし、瀬戸内海と高知の海は同じ海でもまったく違っていたこともよく覚えています。乗り物の旅を通して風土を見る目が養われてゆきました。

中学生になると、電車に乗ってひとり旅へ出かけ、色々な風景や建築を見ることが好きでした。例えば、名古屋から近鉄に乗って、途中下車をしながら、東大寺や室生寺、伊勢神宮へ行ったこともあります。また、父親の仕事の関係で当時まだ共産圏だったハンガリーの首都ブダペストにしばらく滞在していたことがあります。ブダペストでは、世界で初めての電化された地下鉄が走っていて、その歴史を感じたり、地元の人々が乗り降りしている様子や表情を見て、その国の暮らしや社会背景と乗り物との関係を見つめていました。その後パリに行きますが、そこでの経験は圧倒的でした。こんなに歴史的なものが残っていて、それらが現役で現代の生活を根っこから支え、そのことが町に大きな価値を与えていたからです。パリのメトロの出入り口は、町並みに調和するように味わい深いアール・デコの装飾が施されていることや、

パリから各方面に行く鉄道ターミナル駅の旅情溢れる雰囲気に感銘を受けました。

今思うと"乗り物"が好きというよりは、人々の営為や、風土や文化をつなぐものが好きになっていったのでしょう。家にいる時は地図に架空の路線や航路を描き込んで「こんなつながりがあればいいな」と空想を広げていました。またここに寺や城があれば歴史が豊かに感じられるなあ、町が魅力的になるなあ、と地図に歴史のある建物を勝手に書き込んでいたことを思い出します。私にとっての豊かな未来は、見たこともない新奇な建築が織りなす世界ではなく、時間が蓄積された根のあるものに囲まれた世界だったのです。

そんな幼年期や若い頃の経験が今の建築設計の考え方の核をつくっていたのだと思います。

今、日本全国ほとんどの場所が東京から日帰りできてしまいます。便利な世の中ですが、寂しくもあります。効率を重視するあまりその場所ならではの魅力や情感が失われて、地方の町が画一的、奥行がなくなったように思いますし、高速移動によって移り変わっていく風景は人の情報処理のスピードをはるかに超えています。これからは、人の時間の使い方が徐々に変わるにつれ、乗り物の本当の魅力が見直されてゆかなければなりません。

普段は当たり前すぎて気づくことはできませんが、私たちはそれぞれが住む場所において風土の恵みに囲まれて生活をしています。固有の風景、文化、食べ物、慣習があり、それこそがかけがえのない唯

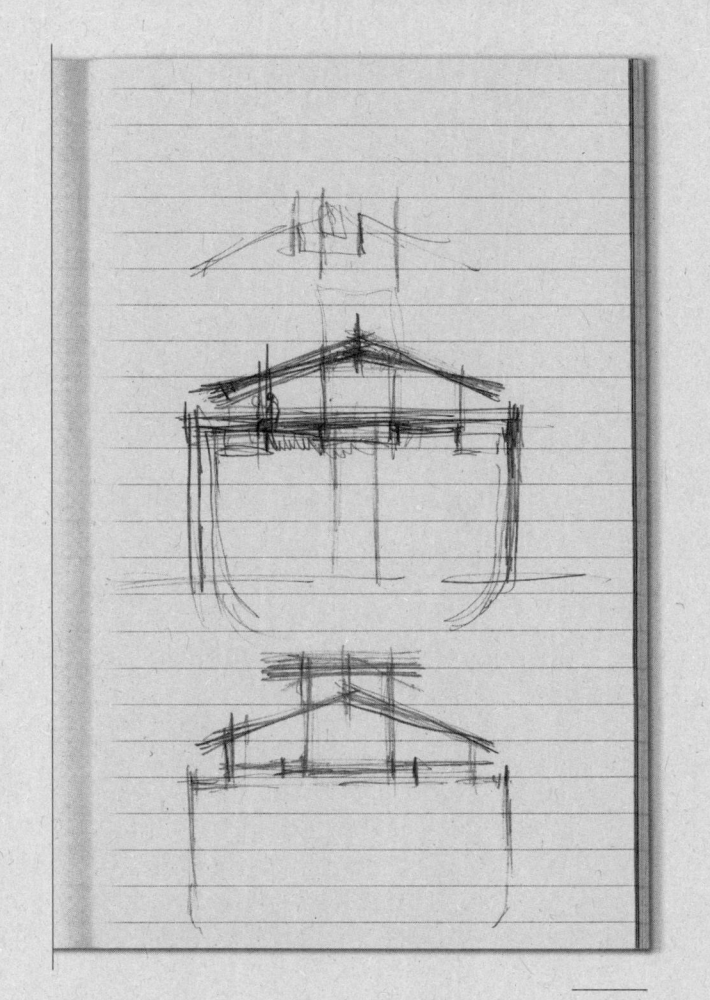

海外滞在中に描いた最初のスケッチ。
縁側があり、切妻屋根の軒が風景を切り取る。

The first sketch I drew on a foreign research trip.
The "engawa" veranda and the eaves of the gabled
roof frame the scenery.

一無二の財産であり、その財産を活かすことがこれからの優れたデザインであると思います。

敷地としての瀬戸内
————

今回、瀬戸内の魅力を掘り起こす客船をつくりたいという設計依頼があってから、その"敷地"となる場所の全体を見るためにヘリコプターに乗って空から視察しました。しかし、空からだけでは人々の営為が伝わってきませんし、スピードも早過ぎて瀬戸内の魅力はわかりません。やはり、瀬戸内を体験するには船が良いということを実感しました。

さて、短い設計期間のなかで瀬戸内の島々や集落の全容を調査することはできません。そこで幼い頃からの記憶、体のなかにあった瀬戸内のトーンを設計の頼りにしました。それは、あえて言葉にするならば、海と空と無数の島々、そして人々の営為が自然に溶け合った"調和"のトーンです。湿度感も独特で、おのずと風景の輪郭はやわらかくぼんやりとします。地中海のように、カラッとしていて、強い光によって物の輪郭がくっきりと英雄的に浮かび上がるような風景とはまったく違います。等身大の何気ない日常と空と海が溶け合っている風景といえばいいでしょうか。

造船のチャレンジ
————

客船「ガンツウ」の設計は、建造する常石造船との二人三脚で進んでゆきました。そもそも日本では、船

舶法および関連法令において平水区域での長時間の宿泊型航海が想定されていなかったのです。常石造船に問い合わせ、国土交通省に確認してもらいながら進めていきました。常石造船としても客船は初めての試みでした。

　設計を始める時に、常石造船の設計チームと海外のクルーズ船に乗る視察に行きました。そのクルーズ自体は素晴らしかったのですが、視察中は異国の風景を見ながら瀬戸内ならではの船がどうあるべきかをずっと考えていました。その帰路で、船に切妻屋根を掛けた最初のスケッチを描きました。瀬戸内の風景を屋根の軒先で切り取ることができたらさぞかし素晴らしいだろうと考えたからです。その船の断面スケッチ[▶P.045]を設計チームに見せて、製造的な問題や、運航上の問題がないかを確認すると、意外にも実現可能であることが見えてきました。一気に船のデザインが切り開かれ、具体的な諸室の配置の検討に移りました。

　まずは操舵室の位置を考えました。船首の上部には最高の部屋をつくりたいという思いがあり、そのために操舵室を1階に下ろしたのです。一番の問題は操舵室からの視界の確保で、その確認には時間がかかりましたが、最終的には許可されました。

　ガンツウの大きさは、乗客・乗員数、モノコック構造による構造力学的な制約、電気推進システムなどの動力的な性能、そして、尾道水道や音戸の瀬戸など瀬戸内海の難所を通ること、橋の下を潜ることなど、様々な条件を満たすものとして導き出されて

いきます。地上3階・地下1階という階数や全長・全幅などおおよその輪郭が依頼前から詰められていたので、素直にそのなかでデザインを進めていきました。

　なんとか期限内に基本プランをまとめることができたのは、普段から住宅の設計をしていたのと、国内外にある魅力的な旅館やホテルに数多く泊まった経験があったからだと思います。等身大のスケール感と複雑な人の動きや心理を学ぶには住宅の設計が最も適していて、初めて手がける客船でもやはり住宅で学んだことが十二分に活かされたのだと思います。また宿泊施設として特に参考にしたのは、歴史的な城や修道院などを改築してホテルとして生まれ変わらせたポルトガルの国営ホテル「ポウサダ」です。時間の蓄積を感じさせ、かつさりげない上質さと素朴さをもった空気感をガンツウのイメージと重ねました。

　2015年の12月、基本設計をまとめて今回のプロジェクトチームにプレゼンテーションを行いました。コンセプトや船のあり方について大きく共感いただき、ほぼそのままの状態で次の段階に進むことになりました。

船のつくり方、建築のつくり方、スペシャリストたちとの協働

————

建築の場合は、地面に通り芯（構造体の中心を表す基準線）を描き、それを基準に構造となる柱や壁などを配置し、下から順に積み上げ、仕上げていきますが、

大名庭園・栗林公園内の茶屋「掬月亭」。

Kikugetsutei, a feudal lord's teahouse in Ritsurin Garden in Takamatsu.

奈良の慈光院の書院。

The reception hall of
Jikoin Temple in Nara.

船の場合は鉄板による構造で、別々につくられた
パーツを最後に一気にドッキングしてつくるので、製
造過程がまったく違っています。通り芯の考え方が使
えないので、同じ間取りの部屋をつくりたくても、そ
れぞれに微妙な調整が必要になりました。陸の建築
に慣れた私たちも、図面を描きながら、だんだんとど
こを基準にするべきかがわからなくなっていくように
感じました。

　躯体ができてから、ドック内で艤装工事という
内装などの工事に入っていきますが、そこでは水平垂
直をどう出すか、という問題もありました。躯体の床版
には水勾配が付けられているので、何を水平垂直と見
なすかが難しく、もちろん建築工事で寸法の押さえ方
の要となる水準器も意味をなしません。そうしたなか
で、唯一フラットに近いのが天井面だったので、そこ
を基準にしてなんとなくつくっていくことにしました。

　手すりのガラスや、客室の窓ガラスの工事は困
難を極めました。通常の船舶では応力の関係から角
が丸い開口になりますが、ガンツウでは建築で使わ
れている矩形のサッシを加工して使っています。揺れ
への対策や水密性などが問題になるため、陸の職人
がすべての責任を取ってやるのは難しく、造船側の
技術者や鋼製建具職人と一緒に検討をし、船なら
ではの特殊なディテールになっています。その場で組
み立てていく建築技術と、厳しい外部環境に対して
求められる性能を担保する造船技術を寄り添わせ、
工夫して展開させていくこと、つまり互いの"いいとこ
取り"によってつくり上げることができました。

自然と制限のなかで
——風、波、重さ、火災
———

構造は造船と建築の両方の観点から検討しました。建築と違って船には地震力は働きませんが、常時の波の揺れ、波荷重を想定しなければいけません。造船の技術は、水に浮かんで停泊している時や動いている時に、いかに全体が動いていない状態をつくるかが重要です。

　屋根は特に風の力を受けるのですが、それに従って船体が動きます。つまり、建築のようにじっとその場で耐えるのとは違うのです。他方で、窓ガラスは建築と同じような風に耐えるための計算をしています。

　ガンツウの仕上げを考えていく段階では、重量がかなり厳しい条件になり、とにかく軽量化が求められました。仕上材だけではなく、例えば浴場の石の下地となるモルタルの重さなども大きく関係してきます。建築も土地や地盤の形質に対応して、ある程度の軽量化を求められる場合がありますが、船のように前後左右の重量バランスまでは重視されることはほとんどないといっていいでしょう。

　外部の塗装は船舶用の専用塗料を使っています。今回ガンツウの色は色自体が主張するものではなく、周囲の風景の色や光を映すような銀色にしています。もっともさりげない色であるように考えた結果です。ちなみに、船の塗装は建築と比べて何倍も厚く塗ります。

重量の制限のほか、防火のルールもとても厳しく、気を配ったところです。一番の難所は、1階と2階の廊下や階段ホールで、そこを通ってオープンデッキに避難するという避難計画になっていますから、不燃材料しか使えません。客室はその次に厳しいところで、普通に考えれば木を使うことなどできないのですが、越井木材という木材の不燃処理を手がけている会社と、何種類もの木材で性能を検証する実験を重ねました。なかなか良い結果が出ず、最後の最後にうまくいったのがアルダーという木でした。幸運なことに、不燃処理した後の色合いがまるでチークのような雰囲気になったので、みんなで喜び合いました。工程としては、このアルダーがもしうまくいかなければ、客室の内装に木を使うことを諦めなければいけないというタイミングでした。重量の制限を考えても、不燃処理さえできれば軽い木は合理的なのです。

　最上階の3階は、建造の経験から見えてくる技術基準を聞きつつ提案をしていきました。その結果、壁や天井・軒裏にサワラ（不燃処理なし）が使えました。サワラは、スギより固く、ヒノキよりは柔らかい材です。スギだと色味の幅があってコントロールが難しくなりますし、ヒノキだと色味は統一できますが、全面に張るとよそよそしく硬い感じが出てしまいます。これまで住宅を設計してきた経験からも、ガンツウにはもう少し温かみと人肌への馴染みの良さがあった方が良いと思いました。サワラは軽量で水に強く、また色味としても目に優しい材です。

奇跡的な調和
———

造船の工程が終わり、2017年1月に初めて船を水に触れさせる進水式がありました。この時それぞれのパーツが合体した様子を目の当たりにします。その時の感動は今でも忘れません。造船の技術、力強さ、スケールの大きさ、そして寛容さのようなものを全身で感じ、しばらく鳥肌が治らなかったことを思い出します。

その後、ようやくドックに入ってきて、ガラス工事などがあり、内装の工事に入れるようになったのが4月です。そこから5カ月ほどですべてを仕上げなければいけないという超突貫工事でした。陸の大工は普段とはまったく異なるつくり方で本当に大変だったと思います。躯体の歪みなどを吸収して美しく見えるように根気強く仕上げてくれましたし、家具・建具職人も、誤差のある造船の躯体の中で的確な判断をし、漆喰を塗る左官職人もオーケストラのような動きで一気に仕上げてくれました。現場では、一時、音や振動への懸念が出たこともありますが、防音・防振の専門家にアドバイスをもらい、天井と床の懐で的確な対策を講じることができました。陸の職人の手配や指揮は地元・福山の大和建設の尽力によるところがとても大きかったと思います。綿密かつ正確な采配によって、奇跡的に期限内に工事を終えることができました。誰ひとり欠けていてもこの船はできていなかったと思います。

ガンツウにとっては夜もとても重要な時間です。その照明計画は、照明デザイナーの面出薫さんのチームに協力を依頼しました。揺れがあるため天井からぶら下がるような器具は使えません。天井に埋め込まれたダウンライトを基本にして、シンプルで素朴な照明計画を、最新の技術を駆使して成立させています。

ともかく関係者すべての専門知識を出し合い、それまでの経験を活かし、総力戦でひとつのものをつくり上げることができたのです。

大変な工事のすべてが終わってから気がついたのは、外構工事がないことです。庭がありません。その代わりに極上の"動く借景"があるのです！

建築と船は共通している
———

ガンツウの設計でなによりも優先したことは、瀬戸内独自の世界観を脚色することなく、あるがままに見せることです。料理でいえば素材の良さをそのまま引き出すことに近いかもしれません。なにかをこねくり回したような創作や演出、あるいはこれ見よがしなデザインはまったく不要なのです。大切にしたのは、風雪に耐えてきた風土、歴史、伝統、風景、人の営為であって、それらを等身大に脚色なく表すことなのです。

龍安寺石庭や大徳寺高桐院、奈良の慈光院など、庭を海に見立てた空間というのは日本に沢山あります。大徳寺孤篷庵や栗林公園の掬月亭などは船そのものがモチーフになっています。人工物と自然をどう調和させるか、そこにある風景に対していかに受

動的（パッシブ）に捉えるかを考えると、やはり屋根や軒、縁が経験的にすんなりと出てきました。しかも風景を切り取る素材は、瀬戸内の穏やかなトーンを考えると木以外には考えられませんでした。

　また海は、災害や塩害をもたらすネガティブな側面もあるため、シェルターであることを基本にして考えなければいけないと思っていました。シェルターとしての安心感があることで、初めて海と人間の良好な関係を築くことができます。大きな屋根に包まれた暮らしは私たちが安心する場所として太古からずっと記憶に刻み込まれています。そんな安心の記憶をガンツウの屋根に期待したのです。

　製造過程や規制は異なりますが、実は建築と船は本来非常に親和性があります。シンプルな構造美と、虚飾がないバランスのとれた意匠の建築を見ると「まるで船のようだ」と感じることが多々あるのです。

　さらに日本建築は親水性ももっているので、船も建築も基本的につくる精神は共通しているといっても過言ではないように思います。嚴島神社や金閣寺を思い浮かべればわかりやすいかもしれません。それこそが初めての船の設計だったにもかかわらず、すっと自然に設計に入り込めた理由だったのではないかと思います。

ガンツウの時間

———

ガンツウは、日本の伝統的な建築がもっている縁側のように、なにもしない時間の豊かさを感じられるよ

龍安寺の石庭。

The rock garden at
Ryoanji Temple in Kyoto.

うな空間があります。ガンツウの縁側や客室、ラウンジでは靴を脱いでもらうようになっています。客室の床仕上げはタモで、素足の気持ち良さと滑りにくさを考慮し、木目を際立たせた浮造り仕上げにしています。縁側には、昔の寺院建築によく使われていたヒノキの厚板を、手に触れる部分には木や籐といった自然素材を用いています。この船でしばらく時間を過ごしてみると、独特の感触が身体に伝わってゆきます。加えて、これら自然素材は経年とともにどんどん魅力や味わいが出てきます。

また、それぞれの空間の重心や、プロポーション、人の目線の高さと軒先の寸法など、身体との関係をとても大事にしました。

さらに大切にしたのはストレスのない動線計画です。人の動きが淀まずに、気持ち良く目的の場所に誘われるような動線を目指しました。“目に見えるものと目に見えないもの”をどちらも大切にしながら、細かな調整を幾度も繰り返したことがガンツウの設計の醍醐味であったように思います。

実際にガンツウが完成して乗船してみると、本当に驚きました。そこから見る瀬戸内海は私が知っていたものとは全然違っていたのです。まったく揺れがなく、静かで安心感のある船から眺める風景だからかもしれません。船首からの風景は、未知の奥行のある風景が延々とやってくるようで、また船の横からの風景は、まるで長く美しい絵巻物をずっと眺めているようでした。その多島美の風景が常に人々の日常の営為とともにあるという環境は、世界のどこにも

類を見ないものではないかと思います。ガンツウに乗ることで知られざる瀬戸内がどんどん発見されてゆきました。

同時に、私自身の五感がどんどん研ぎ澄まされてゆくような感覚もありました。言い換えるならば、現代の生活で眠っていた身体感覚がどんどん目覚めてゆくように感じたのです。刻々と変化する風景や時間の流れを愛おしく感じる感覚。心地良い瀬戸内の風を感じる感覚。建築に使われた素材に加えてリネンやアメニティに用いられた素材の上質さがもたらす感覚。日の出や日の入がこれほど美しく尊いと感じる感覚。そして風土とともにある食を全身で楽しむ感覚……。そんな自分の身体感覚が目覚めてゆくスピードと船の航行のスピードとがピタリと重なり、“ここに確かに自分の心身が存在する”ということを認識できるのです。ガンツウは、効率的な移動手段ではなく瀬戸内独自の世界観と自分の心身を感じるための乗り物であり、建築であり、確かな人の居場所なのです。

設計中に描いた内観パース。
Interior perspectives drawn
during the design process.

Uncovering the Allure of Setouchi

Yasushi Horibe

Memories of travel and transport

——

I design buildings for a living, but my first passion was drawing different means of transport. As a young boy, I lived near the port of Yokohama in Tsurumi, a place where railroads and ships were a familiar part of daily life. From the window of my grandmother's house where I grew up, I could see the ships plying the harbor and the trains running back and forth between Tokyo and Yokohama. I was always sketching ships and trains that I saw, and I delighted in riding them as well. In particular, I fondly recall the month-long summer vacations that our family would take each year to my great-grandmother's home in Kochi Prefecture. Rather than the bullet train, we would travel by the overnight "Blue Train" to Uno in Okayama, then cross by ferry to Takamatsu, where we would board a diesel train

for the final segment of this long journey by rail and sea. The ferry and Blue Train were imbued with a romantic sense of travel that is rarely found anymore. One perk of the journey was enjoying sanuki udon and the Inland Sea breeze on the deck of the ferry on the morning after departure. Food tastes twice as good when enjoyed with a beautiful view and fresh air. This was my first encounter with Setouchi, and the transport, scenery and food remains united in my mind as an indelible original impression. From the ferry I took in the impressive view of the constant ship traffic and countless islands and remember how different the Inland Sea felt from the ocean in Kochi. Travel aboard trains and ships cultivated my eye for the natural features and culture of local places.

When I was a middle school student, I enjoyed taking solo train trips to see various scenery and architecture. I once rode the Kintetsu Line from Nagoya, disembarking to visit the temples of Todaiji and Muroji and the Grand Shrine at Ise. My father's work also took our family to Budapest for a time, when Hungary was still part of the Eastern Bloc. There I rode the world's first electric subway, perceiving how the country's daily life and social system related to its transportation as I absorbed the history and observed the appearance and expressions of the local passengers. Later, I was overwhelmed by my experience of Paris. The city was enriched by so much historical heritage that remained an integral part of modern life. The tasteful art deco entrances to the metro blended seamlessly into the Paris cityscape, and I was entranced by the romance of travel at the railway terminals from which trains depart the city in every direction.

When I think about it now, I came to love the human activity and connections between different locales and cultures even more than the ships and trains themselves. At home, I would draw imaginary rail lines and sea routes onto maps, conjuring up connections that might be. Sometimes I would also add historical buildings, imagining how a temple or a castle in a certain place would make a city more historic or appealing. My idea of a rich future was not a world filled with unfamiliar, novel buildings, but a world made up of rooted things accumulated over time.

These childhood and youthful experiences formed the core of my philosophy of architectural design. Nowadays, it is possible to visit nearly anywhere in Japan on a daytrip from Tokyo. The world has become both remarkably convenient and increasingly sterile. Emphasis on efficiency has eroded the inherent charm and feeling of places, as local towns have become indistinguishable and skin-deep. Aboard high-speed transport, the scenery changes far faster than the human ability to process information. As people begin to spend time in different ways, it is time to reconsider the true allure of transport.

We are so accustomed to the places we call home that we often fail to realize how life in each place is shaped by the blessings of local natural features. Distinctive scenery, culture, food, and customs are irreplaceable, unique heritage, and I believe that good design now depends on making use of such endowments.

各部分がつくられ、
最後にそれらを組み立てる。

Each piece is produced
separately and joined together
at the end.

3階の船骨工事。
上下逆さまにして配管工事
などを行う。

Construction of
the skeleton of the third deck.
Pipe fitting is performed
upside down.

Setouchi as site

After I accepted the commission to create a cruise ship that uncovers the allure of Setouchi, I took a helicopter flight to observe the entirety of the "site" from the air. However, the viewpoint was too high to appreciate the activity of people, and the speed too fast to really grasp Setouchi's allure. The flight confirmed to me that the best way to experience Setouchi is aboard a ship.

Of course, it was not possible in the short design period to explore each of the islands of Setouchi, or the entirety of any particular settlement. I rooted my design in my early memories and the tone of Setouchi I had inside me. If I were to put it into words, this is a tone of harmony between the sea, sky, countless islands, and the human activity that blends into the natural landscape. There is a particular sense of temperature and the outline of the landscape seems to become soft and indistinct. It is completely different from places like the Mediterranean, where the pleasant dryness and strong light makes the scenery appear in bold and clear silhouette. Setouchi is a landscape in which life-sized everyday activity blends together with the sky and the sea.

The challenge of shipbuilding

The design process for guntû progressed hand-in-hand with the shipbuilder, Tsuneishi Shipbuilding. To begin, the Law on Ships and other Japanese laws and regulations did not foresee the operation of extended passenger cruises with overnight accommodation in the placid waters of a protected sea. We were in constant contact with Tsuneishi Shipbuilding and the Ministry of Land, Infrastructure, Transport and Tourism as we moved forward with design work. Guntû was also Tsuneishi Shipbuilding's first time building a passenger ship.

At the start of the design process, I went on a research trip to ride an overseas cruise ship along with the Tsuneishi Shipbuilding design team. The cruise itself was a wonderful experience, but even as we took in the exotic scenery during the journey, I was constantly thinking about how to create a ship that would uniquely reflect Setouchi. On the way home, I made my first sketch of a ship with a gabled roof. I thought that the gabled roof would surely provide a spectacular frame for the scenery of Setouchi. I showed the cross-sectional sketches of the ship [▸ P.045] to the design team and asked about challenges related to construction and operation, and learned that the concept was surprisingly feasible. The design began to unfold all at once, and we moved on to thinking about the exact placement of the rooms.

First, we considered the location of the bridge. I wanted the most luxurious suite to occupy the bow of the ship on the upper deck, so the bridge was moved down to the first deck. The biggest challenge was ensuring visibility from the bridge, but after some time we eventually received approval.

Guntû's size was determined according to various conditions, such as the number of passengers and crew, limitations of structural strength resulting from the monocoque structure, the output of the electric propulsion system, as well as the ability to navigate through narrow sections of the sea such as the Onomichi Channel and Ondo-no-seto Strait and pass beneath bridges. The basic specifications of the ship, such as three above-water decks and one below-water deck and the total length and width, had mostly been decided before I was commissioned, so the design proceeded within those parameters.

My background designing homes and personal experience visiting many appealing ryokan and hotels across Japan and overseas was crucial for pulling together the basic plan within the deadline. Designing homes is the best way to develop a life-sized sense of scale and learn the complexity of how people move and think. That experience proved exceptionally useful for my first ship design. Inspiration also came from the Pousadas run by the Portuguese government that transform castles, monasteries, and other historic buildings into hotels. I folded this sense of accumulated history and unpretentious elegance and simplicity into my imagination of guntû.

In December 2015, I wrapped up the basic design and presented to the project team. The reaction to the concept and character of the ship was very positive, so we proceeded to the next stage with few changes.

Ways of shipbuilding, ways of architecture, and working with specialists

——————

In architecture, we draw a base line on the ground that represents the center of the structure, and use that as a reference for placing the structural columns and walls, and then add layers from the bottom up until complete. However, ships are made from steel plates and built using an entirely different production process in which separately made parts are pieced together at the end. Here the notion of a base line is useless, so even when we wanted to create two rooms with identical layouts, each required minor adjustment. Accustomed as we were to building on land, we would gradually become uncertain of what to use as a basis for drawing our plans.

Once the hull is completed, shipbuilding moves on to the stage of construction known as outfitting when interior furnishings and other components are installed inside the dock. During this stage, we needed to figure out how to determine the horizontal and vertical planes. The floor of the hull has a gradient for water drainage, so it is unclear what can be considered horizontal and vertical. Levels, which are essential tools for determining measurements in building construction, are meaningless. In this case, we found that the only surfaces that were close to flat were the ceilings, so we used them as the basis and built out the spaces from there.

Making the railing glass and window glass in the passenger cabins was extremely challenging. In most ships, stress considerations dictate that the corners of windows are rounded, but for guntû we altered rectangular frames that are ordinarily used in buildings. Craftsmen who usually work onshore had to consider special ship-specific details together with the shipbuilding engineers and metal fitters to ensure that the fixtures could withstand swaying and water

2017年1月16日進水式。

The ceremonial ship
launching was held on
January 16, 2017.

exposure. We were able to complete the ship by creatively bringing together architectural techniques suited to constructing something on site and shipbuilding technology that ensures performance necessary to withstand the harsh external environment—in other words taking the best parts from both worlds.

Working within nature and limitations: wind, waves, weight, and fire

———

We considered the structure from both shipbuilding and architectural perspectives. While ships are not impacted by earthquakes like buildings, the constant swaying and weight of waves must be assumed. Shipbuilding technology emphasizes creating a stable onboard environment both when the ship is anchored in water or underway.

The roof in particular absorbs the brunt of the wind, but guntû can move in response, unlike buildings that must simply withstand forces. Even so, window glass is calculated to endure the same wind strength as buildings.

When considering the finish materials on guntû, it was important to reduce weight as much as possible. This meant factoring in not only the finish materials themselves, but also the weight of materials such as the cement used underneath the stone in the bathrooms. Sometimes the characteristics of the land or soil make it necessary to reduce the weight of a building somewhat, but rarely if ever do architects need to focus on balancing weight between front and back, or left and right.

The exterior is covered in a special type of paint used for ship hulls. Guntû's color is not assertive, but rather a silver that reflects the hue and light of the surrounding scenery. It was the most unassuming color I could think of. Ships are actually covered in paint many times thicker than buildings.

In addition to weight limitations, we also needed to think hard about how to meet the strict rules for fire prevention. The most challenging areas were the corridors on the first and second decks and the stairwell, which form the evacuation route for passengers to escape onto the open deck and could therefore only use non-flammable materials. The passenger cabins were the next most difficult places and ordinarily could not use wood. However, we partnered with a company called Koshii Woods that processes wood to be inflammable and conducted experiments with a number of different types to verify their performance. We struggled to get good results, but at the very end a wood called alder finally showed promise. We were thrilled that after processing the shade ended up resembling teak. Alder was the final species that we could test before the construction schedule would have forced us to abandon hope of using wood in the interior decoration of the cabins. Considering the weight limitations, light wood was a rational choice if only we could make it inflammable.

I came up with proposals for the third and highest deck as I inquired about technical standards derived from prior shipbuilding experience. As a result, we were able to use sawara cypress (without treatment for inflammability) for the walls, ceiling, and beneath the eaves.

進水後、内装工事に入る。

Beginning the interior construction after the ship launching.

木をふんだんに使った内装工事。

The interior design made abundant use of wood.

Sawara is harder than cedar and softer than hinoki cypress. It is difficult to control the tone variation of cedar and although the shade of hinoki can be controlled, using it everywhere results in a cold, unfriendly mood. From my experience designing homes, I felt that guntû needed something with a little more warmth and affinity for human touch. Sawara is light and water-resistant and also has a nice tone that is pleasing to the eyes.

Miraculous harmony

After the hull was finished, the ship first entered water in the launching ceremony in January 2017. At this moment, all the separate parts suddenly become a whole. Even now, I cannot forget how moved I was. I felt with my whole body the technology, forcefulness, magnitude, and even tolerance of shipbuilding. My excitement did not abate for some time afterward.

It was April by the time the ship entered the dock and installation of windows and interior finishes began. From there, it was an all-out race to finish everything in a span of just five months. The carpenters were put to the test by the challenge of working in an entirely different way than they would on land. They tirelessly applied the finishing touches that make the ship shine and absorbed the imperfections of the hull. At the same time, the furniture makers and fitters made apt decisions in order to accommodate minor errors in hull construction and the plaster-ers moved like an orchestra to smoothly finish the walls. When concerns arose about noise and vibration, we incorporated advice from mitigation experts and took effective measures within the ceilings and floors. Essential to the project was the commitment of Daiwa Construction, the local Fukuyama contractor that mobilized and supervised the carpenters. Thorough and precise supervision made it possible to miraculously complete construction on schedule. The ship could not have been completed without every single person involved.

Night is a very important time on guntû. We asked lighting designer Kaoru Mende's team to prepare the lighting plan. The swaying of the ship made fixtures that hang from the ceiling unfeasible. We used the latest technology to realize a simple and modest lighting plan that is centered around downlights that are embedded in the ceiling.

Everyone involved brought specialized knowledge and experience to the table, joining forces to create something together.

After all of the challenging construction was over, I realized there was no outdoor landscaping to finish. There is no garden. Instead, the ship has the most spectacular moving backdrop!

Commonalities between buildings and ships

———

My main concern was that I did not want guntû's design to overdramatize the distinctive world of Setouchi, but to simply show it as it is. To make a food analogy, this is akin to bringing out the essential flavor of ingredients. It is completely unnecessary to contrive a distracting form or performance or an ostentatious design. I was focused on the local natural features, history, tradition, scenery, and human activity that has endured over the centuries, and the task of presenting these at a life-sized scale without embellishment.

There are many spaces in Japan where gardens are used to symbolize the sea, such as the rock garden at Ryoanji Temple, Koto-in at Daitokuji Temple in Kyoto, or Nara's Jikoin Temple. Ships themselves become motifs at Daitokuji Koho-an and Kikugetsutei Teahouse in Ritsurin Garden in Takamatsu. My experience led me to use the roof, eaves, and edges as a means for harmonizing manmade and natural elements and passively relating to the scenery outside. Considering Setouchi's mild tone, I couldn't imagine using any material other than wood to frame the landscape.

The sea also has a dark side as the source of disasters and salt pollution, and thus I felt that I had to think about the structure first and foremost as a shelter. The feeling of safety inside a shelter is vital for people to cultivate a positive relationship with the sea. Since ancient times, we have instinctually felt safe beneath large roofs. I hoped that guntû's roof would evoke that memory of safety.

Although the production process and regulations differ, there is in fact much in common between buildings and ships. We often remark that a building's simple structural beauty or well-balanced, unostentatious design resembles a ship.

Japanese architecture also has an affinity for

water, so it is not a stretch to suggest that there is a common mentality behind making ships and buildings. Examples that come to mind are Itsukushima Shrine and Kinkakuji Temple. Perhaps this is why the design came naturally to me even though guntû was my first ship.

Time on guntû

Guntû is a space where one can enjoy the luxury of doing nothing, much like the "engawa" veranda of traditional Japanese buildings. The verandas, cabins, and lounge on guntû are made to be enjoyed barefoot. The floors of the cabins are finished in ash, which was chosen for its barefoot comfort and non-slip qualities and is processed to bring out the textured grain of the wood. The verandas are finished with thick blocks of hinoki, frequently seen in old temple architecture, and use natural materials such as wood or cane for areas that come into contact with skin. After spending a while on board, the body absorbs the distinctive texture of the ship. Moreover, these natural materials become ever more charming and tasteful as they age.

Each element was considered in the context of how it relates to the body, from the balance and proportions to the height of people's eyes or the dimensions of the eaves in each space.

I also took care to design a stress-free pattern of circulation. I tried to create paths that would avoid blockages of movement and warmly invite people toward their destinations. The true pleasure of designing guntû was going through countless iterations as I balanced both visible and invisible aspects.

When guntû was finished and I first stayed on board, I was truly surprised. Seen from guntû, Setouchi was completely different from what I had known. Perhaps it was because the ship did not sway and was quiet and stable as I gazed out at the scenery. From the bow of the ship, the view appears as the interminable approach of unknown scenery from deep in the distance, while the vista from the sides resembles the unfurling of a long, beautiful scroll. I think there is nowhere on Earth that quite compares to this environment, where a stunning landscape of countless islands exists alongside everyday activity. Aboard guntû, I came to discover ever more unknown parts of Setouchi.

I also felt that the experience sharpened my own senses. To put it differently, I felt as if I had awakened the bodily senses that had become dormant in modern life. The feeling of fondly sensing the gradual passage of scenery and the flow of time. The feeling of the gentle Setouchi breeze. The feeling of refinement exuded by every material, linen, and amenity. The feeling of heavenly beauty in the sunrise and sunset. The feeling of the whole body savoring food inherent to local ways of life. The pace of my own bodily senses awakening and the speed of the ship's travel seemed to perfectly coincide and I recognized that I was present completely in both body and mind. Guntû is not an efficient means of transport; it is a vessel, a piece of architecture, and safe haven for people to feel the distinctive world of Setouchi along with their own body and soul.

1階の操舵室。

The bridge
on the first deck.

ドローイング
Drawings

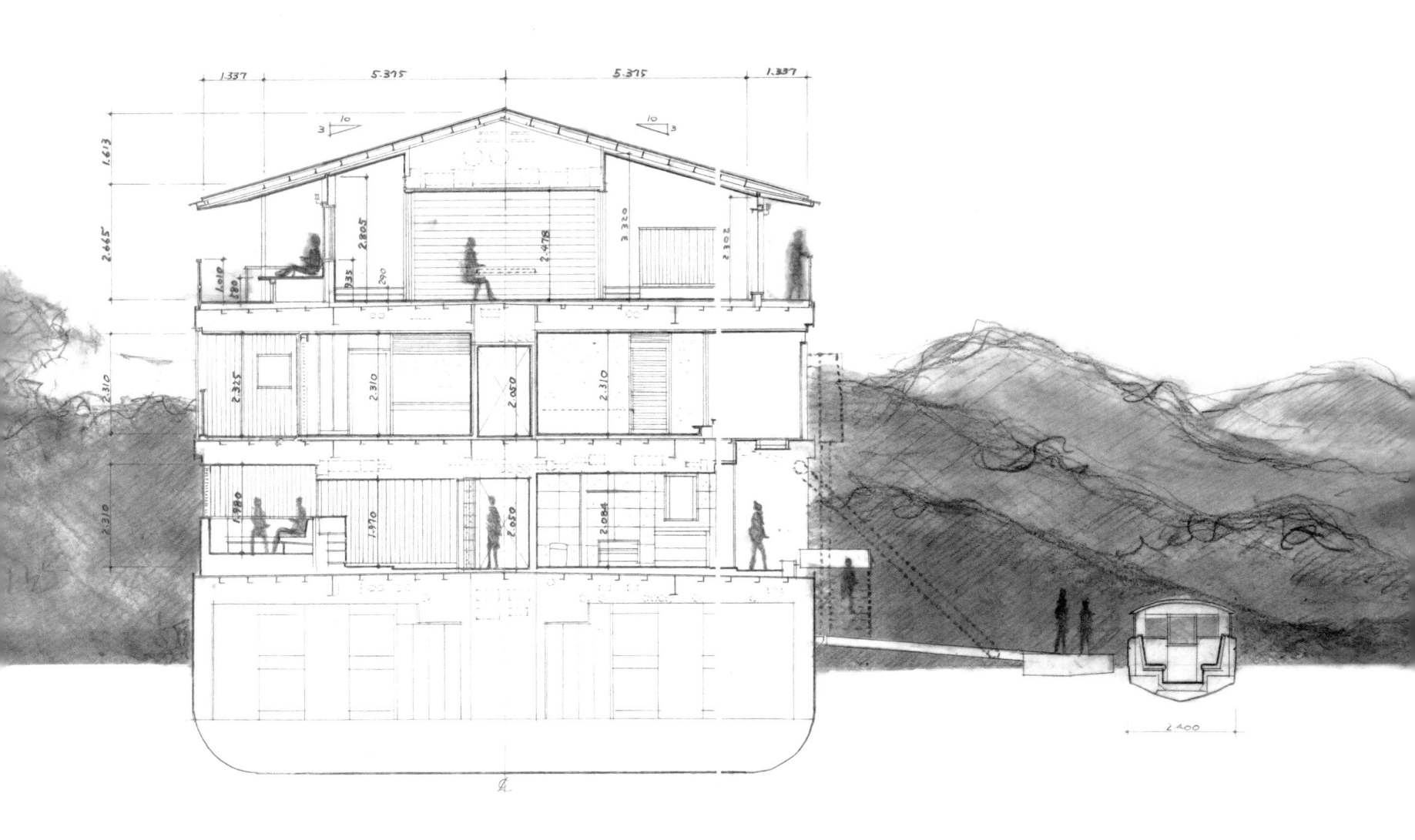

9.260

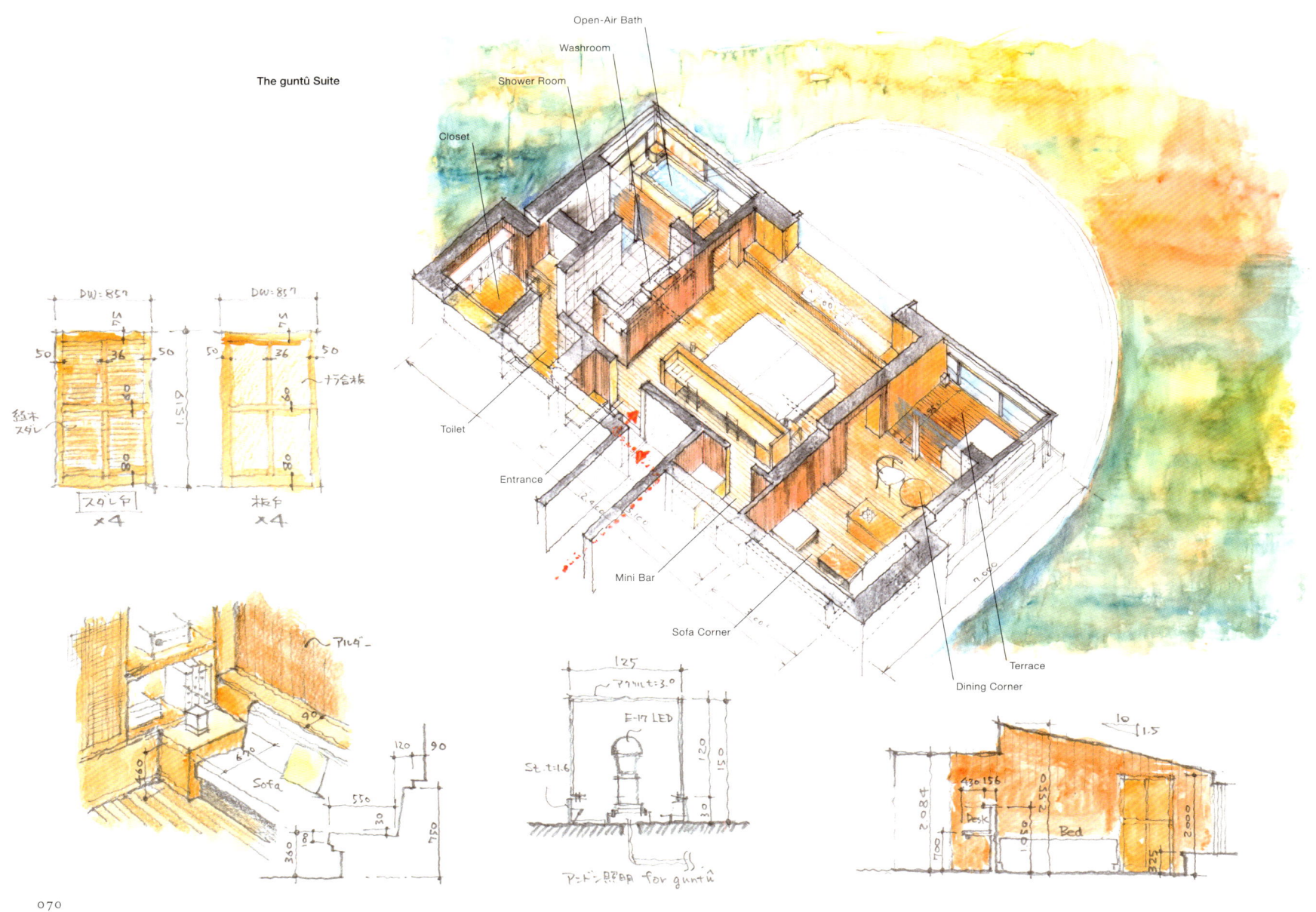

The guntû Suite

Open-Air Bath
Washroom
Shower Room
Closet
Toilet
Entrance
Mini Bar
Sofa Corner
Dining Corner
Terrace

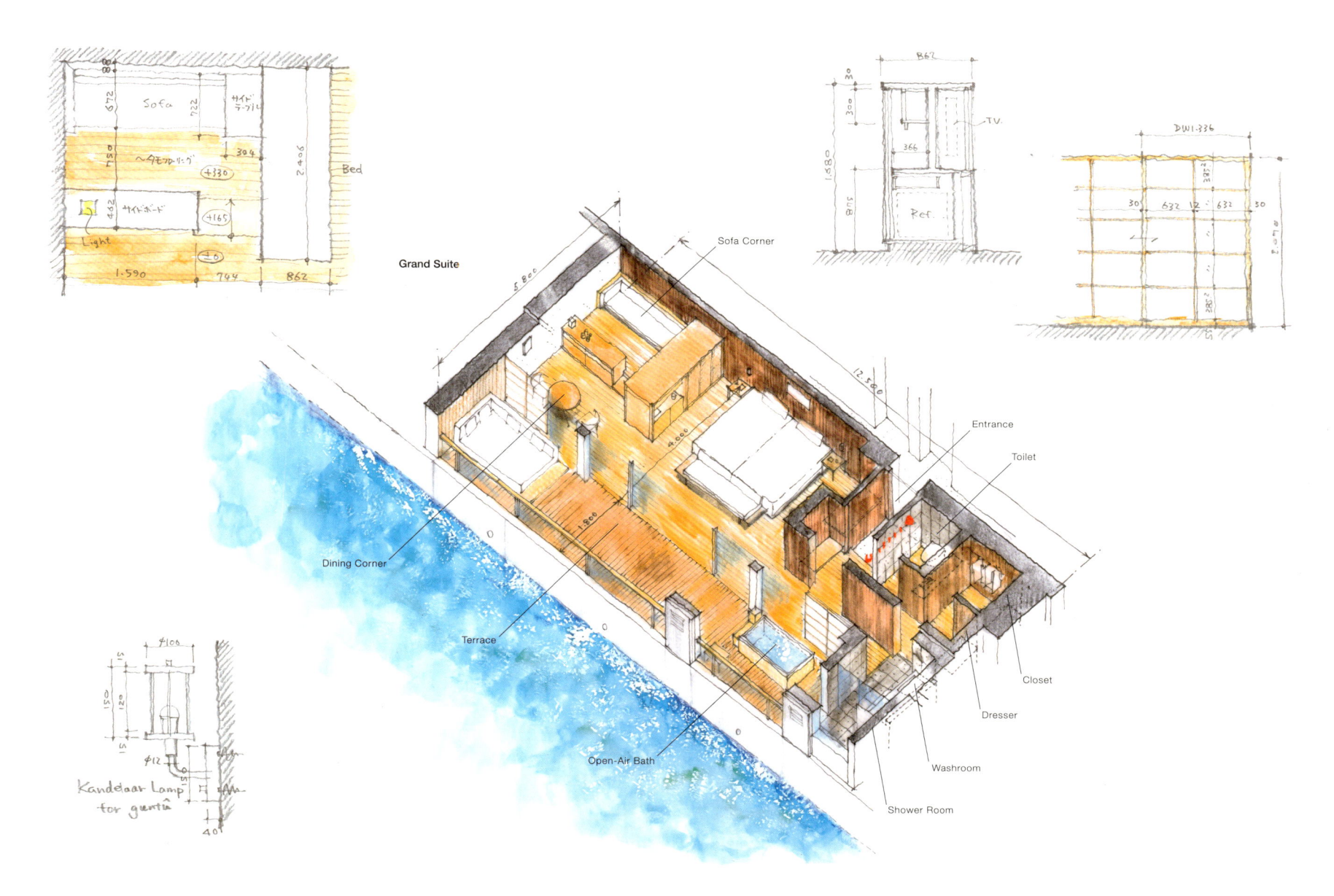

Grand Suite

- Sofa Corner
- Entrance
- Toilet
- Closet
- Dresser
- Washroom
- Shower Room
- Open-Air Bath
- Terrace
- Dining Corner

Kandelaar Lamp for gentie

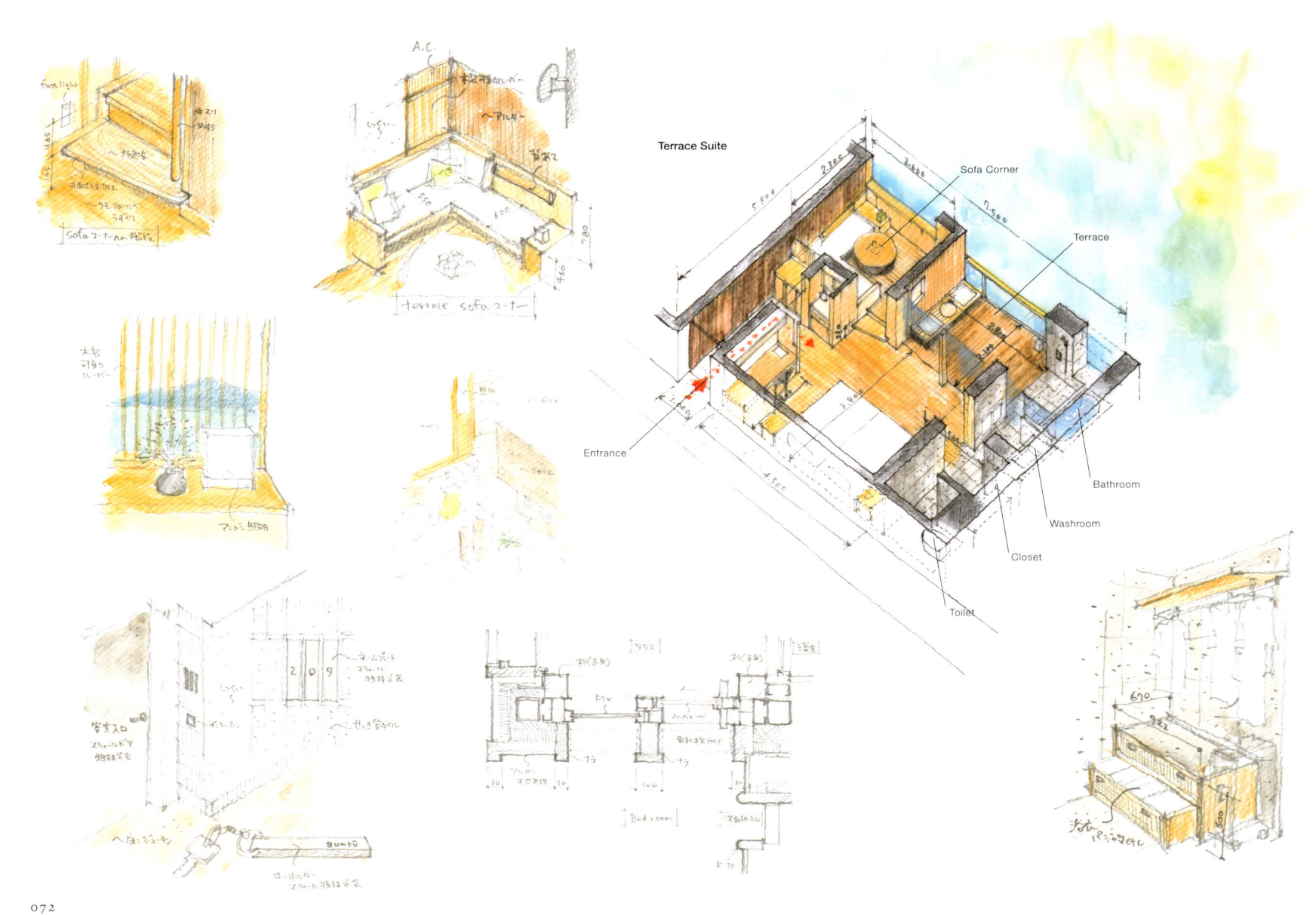

Terrace Suite

Sofa Corner

Terrace

Entrance

Bathroom

Washroom

Closet

Toilet

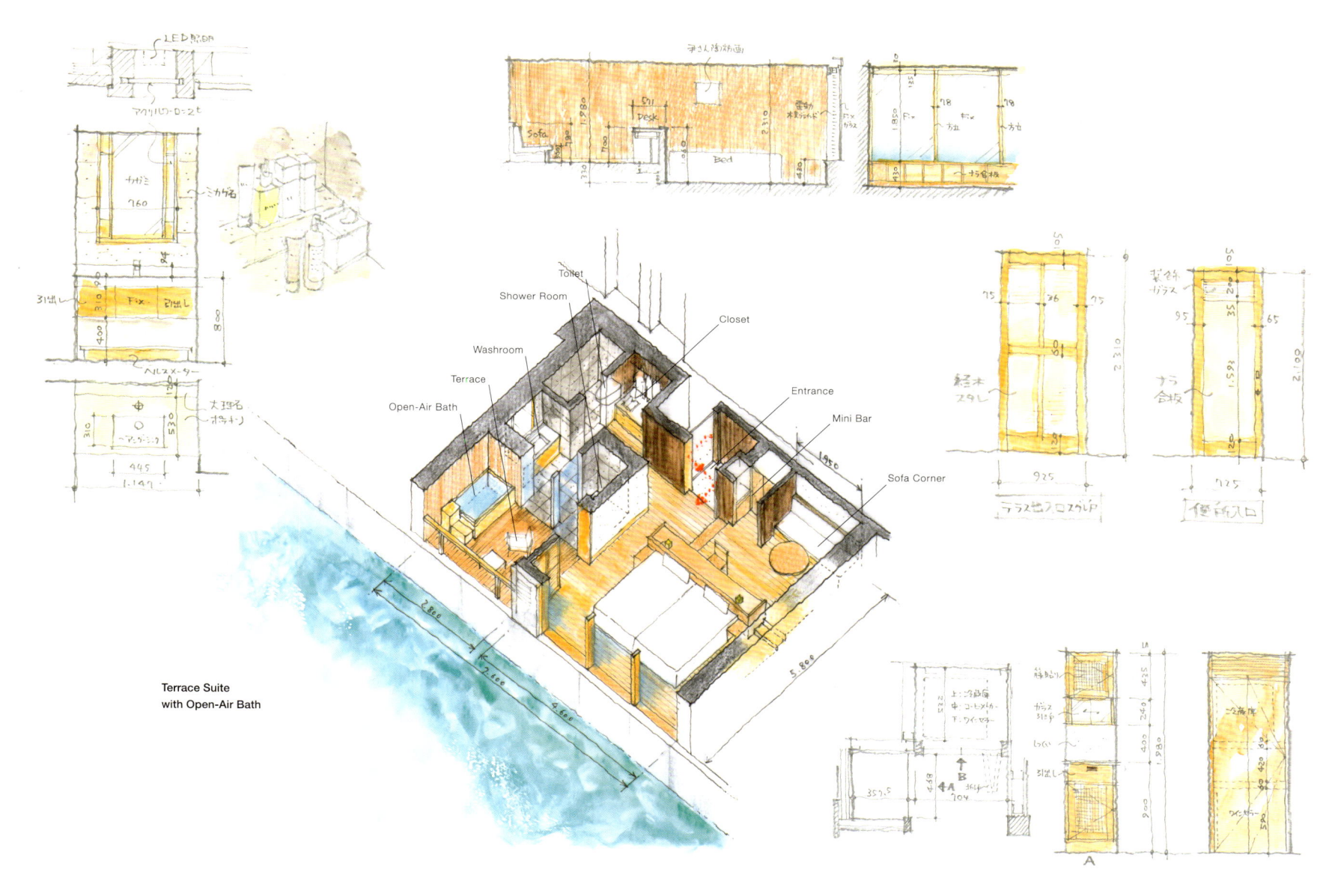

Terrace Suite
with Open-Air Bath

- Toilet
- Shower Room
- Washroom
- Terrace
- Open-Air Bath
- Closet
- Entrance
- Mini Bar
- Sofa Corner

ガンツウの全貌
Overview of guntû

ザ ガンツウスイート
The guntû Suite

2階船首にひとつだけある
客室で、最も広く、約90m²。
窓からは進行方向の景色を
望むことができる。
ふたつのテラスがあり、
左舷側には露天風呂がある。
壁と天井は不燃処理を施した
アルダー。

This suite is the only cabin
facing the bow on the second
deck and the largest at 90
square meters. The windows
offer views of the scenery in
the direction of travel.
There are two terraces that
include an open-air bath on
the port side. The walls and
ceiling are finished with
inflammable alder wood.

グランドスイート
Grand Suite

部屋の1面はすべて開口部で、
広いテラスに連続する。
全体が縁側のような空間（約80m²）。
床は浮造りにしたタモ。

The exterior wall of this cabin
features floor-to-ceiling windows
that open onto a spacious
terrace, giving the entire space
the feel of an engawa veranda.
Around 80 square meters.
The floors are covered in
textured ash wood.

テラススイート

Terrace Suite

海側には寝室から330mm上がった
ソファコーナーがあり、
ゆったりと景色を眺めることができる。
約50m²。

This cabin features a seaside sofa corner, where the floor is elevated 330mm from the bedroom and the scenery can be enjoyed at a relaxed pace. Around 50 square meters.

テラススイート
露天風呂付き

Terrace Suite
with Open-Air Bath

テラススイートのバリエーション。
テラスにヒノキの露天風呂がある。約50m²。
4タイプどの客室も海に面している。
また、瀬戸内で採集した砂でつくった顔料により
描かれた尹煕倉氏の絵画《瀬戸内》が
掛けられている。

This variation of the Terrace Suite
features an open-air hinoki bath on the terrace.
Around 50 square meters.
All four cabin types face the sea.
The suites also feature artwork from the
"Setouchi" series by artist Heechang Yoon
made with pigment produced from sand
gathered in Setouchi.

P.082 [left]

階段の吹抜け上部には切妻屋根が見え、
自身の位置を把握するための指標となる。

The gabled roof can be seen above the
open stairwell, allowing guests to grasp
their location.

P.082 [right]

廊下。階段や廊下は
防火や重量などの厳しい条件により、
漆喰とせっ器質タイルで仕上げている。
漆喰は割れ防止のため特殊な
下地処理を施している。

Corridor. The stairs and corridors are
finished with plaster and stoneware tiles
to conform to rules for fire prevention
and to reduce weight. The plaster was
spread on a special sublayer in order to
prevent cracking.

P.083

1階エントランス。
エントランスサッシの室内側には、
小沢敦志氏による鉄の格子が
設けられている。階段は鉄製。
切妻屋根や木の仕上げとは対比的に、
曲線を用い、漆喰で白く仕上げている。

Entrance on the first deck.
The interior side of the entrance is
decorated with a steel lattice made by
Atsushi Ozawa. The stairs are made of
steel. Contrasting with the gabled roof
and wood finishes, the stairs are curved
and finished with white plaster.

Overview of guntû

3階ダイニング。
床はクリ、壁と天井はサワラによる仕上げ。
窓や間仕切りに、光が調整できる
木製可動ルーバー戸を設けている。
照明は揺れを考慮して
ダウンライトを基本とし、
すべての器具を特注で製作した。
和食の監修は、
東京の「重よし」の佐藤憲三氏。

The dining area on the third deck. The floor is finished with chestnut and the walls and ceiling with sawara cypress. Windows and room dividers are equipped with movable wooden louvers to control light. The lighting design employs custom-made downlight fixtures so as not to be disturbed by the swaying of the ship. Japanese cuisine is overseen by Kenzo Sato of Tokyo's Shigeyoshi restaurant.

鮨。カウンター形式の6席。
監修は「淡路島 互」で、
大将の坂本互生氏にプランを見てもらい、
職人の動きや対応人数などをヒアリングした。
一度カウンター内に入ったら
頻繁には出入りしないと聞き、
設計当初からはカウンターに対する
客席の向きを反転。
ダイニングから独立した
離れ小島のようになった。
ゲストとの関係において重要な
カウンター幅や寸法や素材、
繊細な食材を扱う手元を照らす照明に
気を遣った。

The sushi bar. Guests sit at a six-seat counter. Sushi is supervised by Nobuo Sakamoto of Nobu at Awajishima, who reviewed the plan and gave advice about the movement of chefs and number of guests. After hearing that chefs rarely leave the counter, I changed the original design so that diners sit on the opposite side. The counter forms a sort of island separated from the rest of the dining area. I paid particular attention to factors integral to interactions with guests, such as the width, depth, and materials of the counter and the lighting of delicate ingredients near the hands.

ラウンジ。
3階船尾にあり、靴を脱いで入る。
壁と天井はサワラ、床はじゅうたん。
開口部と外の手摺りとの関係から、
床レベルを400mm上げている。
より切妻屋根が近くなり、
木に包まれたような空間。
立礼のテーブルは、鮨カウンターと同様に
最初は職人の動線を考えて
床の間側にあったが、独立させて
船尾の方へもっていくことで落ち着いた。
和菓子は奈良県「樫舎」
喜多誠一郎氏による監修。

The Lounge. Located at the stern on the third deck, this space is entered after removing shoes. The walls and ceiling are sawara cypress and the floor is carpeted. Due to the position of the windows and the railings outside, the floor is raised by 400mm. As a result, the space is closer to the gabled roof and feels wrapped in wood. Similar to the sushi bar, the counter where tea and sweets are prepared was originally planned on the alcove side of the room out of consideration for the movement of the staff, but was eventually relocated to a separate place at the stern. Japanese sweets are overseen by Joichiro Kita from Nara's Kashiya.

P.090 [left]

3階のカフェ＆バー。U字型のカウンター。

The Café & Bar on the third deck features a U-shaped counter.

P.090 [right]

オープンデッキでの朝食。

Breakfast on the open deck.

P.091

食事は「お好きなものを、お好きなだけ」がコンセプトで、季節に応じた新鮮な地の物を食することができる。船名の「ガンツウ」は尾道地方の方言でイシガニを意味する。良い出汁となる。

Dining is based on the concept of "what you want, as much as you want," with guests enjoying fresh local ingredients specific to each season. "Guntû" comes from the name in the Onomichi dialect for a local blue crab that makes excellent dashi soup stock.

P.092 [left]

2階のトリートメントルーム。
楕円の壁の内側は浴室。

The treatment room on the second deck.
The bathtub is located inside of the round wall.

P.092 [upper right]

2階船尾のサウナ。
最初に自然と人間をつなぐ場所としての
サウナの素晴らしさを知ったのはフィンランドで、
ガンツウにも必須だと思って提案した。
ドライサウナとスチームサウナの2種類がある。
ドライサウナの室温は90度より少し低いくらい、
水風呂の水温は18度。

The sauna at the stern on the second deck.
I first learned in Finland that saunas can be
wonderful places connecting humans with
nature and thought that this would be an essential feature on guntû. There are two types:
a dry sauna and a steam sauna. The temperature
of the dry sauna is slightly below 90 degrees;
the cold water bath is 18 degrees.

P.092 [bottom right], P.093

船の後方を望む浴場。
日毎に男湯と女湯が入れ替わる。浴槽はヒノキ。

The bathing area overlooks the ship's trail.
The two sides alternate daily between men and
women. The bathtub is made of hinoki.

ゲストを「朝さんぽ」などの
アクティビティに案内するための
テンダーボート「ねぶと」。
当初は、切妻の屋根を載せた
オープンなものを提案していたが、
ガンツウと違って
スピードを出す必要があるので、
風の抵抗や室内環境を考え、
開口部を絞った流線型の形を
スケッチしていった。
FRPの3次元曲面によって成形。
アルミ合金で造船された。

The "Nebuto" speedboats are
used to take guests on activities
such as morning walks. I originally proposed an open design
with a gabled roof, but the need
for greater speed than guntû
raised issues with air resistance
and comfort, leading to sketches
of streamlined shapes with few
openings. The boats were
formed with 3-D curved
fiber-reinforced polymer (FRP)
and built from aluminum alloy.

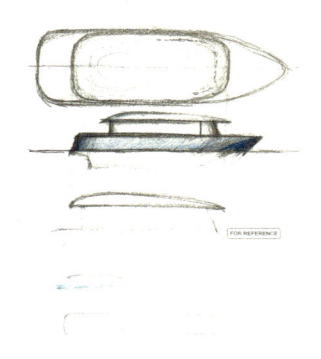

FOR REFERENCE

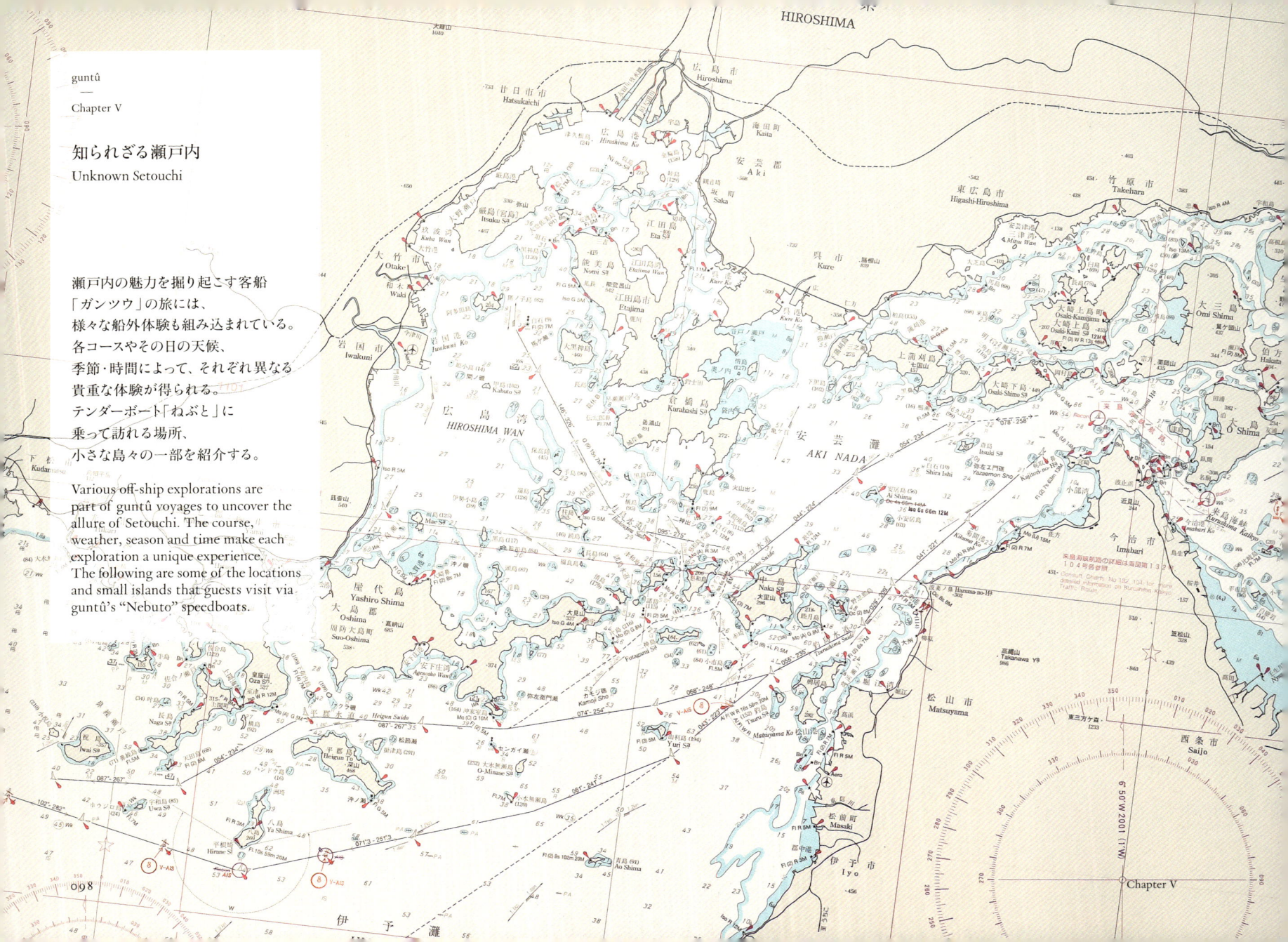

guntû
—
Chapter V

知られざる瀬戸内
Unknown Setouchi

瀬戸内の魅力を掘り起こす客船
「ガンツウ」の旅には、
様々な船外体験も組み込まれている。
各コースやその日の天候、
季節・時間によって、それぞれ異なる
貴重な体験が得られる。
テンダーボート「ねぶと」に
乗って訪れる場所、
小さな島々の一部を紹介する。

Various off-ship explorations are
part of guntû voyages to uncover the
allure of Setouchi. The course,
weather, season and time make each
exploration a unique experience.
The following are some of the locations
and small islands that guests visit via
guntû's "Nebuto" speedboats.

本島 ｜ 香川県丸亀市

Honjima ｜ Marugame, Kagawa

備讃瀬戸に浮かぶ塩飽諸島の中心であり、
海運・廻船業で栄えた
塩飽水軍の本拠地であった。
笠島集落では、船大工の技術を活かした
塩飽大工の仕事を見ることができる。
1985年に
「笠島重要伝統的建造物群保存地区」
に指定され、江戸時代の敷地割りや、
江戸後期から昭和初期にかけての
建築物が良好な状態で残る。

Honjima is the center of the Shiwaku
Islands that are scattered between
Kagawa and Okayama and was the
main base of the Shiwaku pirates, who
flourished as transporters and ship
pilots. The village of Kasashima
features the work of Shiwaku carpenters who made use of techniques from
shipbuilding. Designated as the
Kasashima Historical Preservation
District in 1985, the settlement still
has a layout dating back to the Edo
Period and numerous well-preserved
19th- and early 20th-century buildings.

Unknown Setouchi

103

厳島｜広島県廿日市市
Itsukushima｜Hatsukaichi, Hiroshima

世界遺産・厳島神社を擁する神の島。
通称宮島。
古代より主峰弥山、島自体が
信仰の対象となり、
海上交通の要所でもあった。
厳島神社は、沖に大鳥居があり、
その軸線の先の渚に建つ社殿は、
満潮時、海上に浮かんでいるような
姿を見せる。
多くの観光客で賑わうエリアの裏には
静かな町並みが広がる。

This island of the gods is home to
the World Heritage Itsukushima
Shrine and is commonly known as
Miyajima. The entire island and
its main peak of Mount Misen
were objects of worship and a vital
point for seaborne transport.
Itsukushima Shrine's torii gate is
located in the shallow waters
pointing towards the shrine buildings that stand along the shore.
At high tide the shrine appears to
float on the sea. Beyond the waterfront area bustling with tourists
sits a quiet town.

鹿島｜広島県呉市

Kashima｜Kure, Hiroshima

広島県の最南端に位置する
人口約300人、
半農半漁が営まれている島。
1860年代、幕末の頃に定住が始まり、
山の斜面に石垣による
美しい段々畑がつくられてきた。
ちりめんやひじき、あかもくなどの
生産でも知られている。

Kashima is the southernmost
island in Hiroshima Prefecture,
home to approximately 300 people
engaged in farming and fishing.
Beautiful terraced fields started to
be built on the mountain slopes
after permanent settlement began
near the end of the Edo Period in
the 1860s. The island is also
known for producing sardines and
hijiki and akamoku seaweed.

鞆の浦 | 広島県 福山市

Tomonoura | Fukuyama, Hiroshima

古くから潮待ちの港として栄え、
今も趣深い町並みが残る
瀬戸内を代表する景勝地のひとつ。
海際には江戸の
港湾施設だったことを示す
常夜燈、雁木などの遺構がある。
沖には仙酔島や弁天島、
皇后島などの小島が浮かぶ。

The port of Tomonoura prospered since ancient times as a harbor to wait for the tides and retains an atmospheric cityscape that is famed as one of Setouchi's most beautiful locations. The Joyato Lighthouse that signaled the entry to the port during the Edo Period stands along the waterfront near the remains of stone steps leading into the harbor. Just off the coast are minor islands such as Sensuijima and Bentenjima.

祝島 | 山口県熊毛郡

Iwaishima | Kumage District, Yamaguchi

周防灘と伊予灘の間、
瀬戸内海の海上交通の
要衝に位置する。
主に漁業と有機農業で
生計が立てられている。
石積みの棚田や、
練った土と石を交互に積み上げ
表面を漆喰で固めた
「石積み練塀」が見られる。

This island sits in between
Yamaguchi and Ehime
prefectures at a strategic location
for transportation within the
Inland Sea. Residents mostly
practice fishing and organic
farming. The village features
stone-walled terraced fields and
stacked stone walls in which
mixed dirt and stones are layered
and hardened with plaster.

北木島 | 岡山県笠岡市

Kitagishima | Kasaoka, Okayama

小さな島ながら、
瀬戸内海中部の笠岡諸島では最大。
江戸時代より石の島として栄え、
良質の花崗岩は「北木石」という
ブランドになっている。
島内には採石跡が多く残る。
1892年より現在まで
採石が続く鶴田丁場では、
地下深くまで掘り進められた
圧巻の光景に出会う。

Although small, Kitagishima is the
largest of the Kasaoka Islands that
span the central Inland Sea.
The island's stone industry has
prospered since the Edo Period and
its high-quality granite is sold
under the brand Kitagi-ishi.
Numerous quarry remains dot the
island. Excavation at the Tsuruta
Quarry has continued from 1892
until the present, creating a
stunning vista of the deep pit.

ガンツウ guntû					Gross Tonnage	3,013t		Total Floor Area (hotel portion only)

ガンツウ
guntû

総トン数	3,013t
全長	81.2m
全幅	13.75m
主機関	水冷式三相誘導電動機2機
運航速度	10ノット

延床面積（ホテル区画のみ）
2,639m²
1階600m²/2階1,136m²/3階903m²

客室	19室
乗客定員	38名

Gross Tonnage	3,013t
Full Length	81.2m
Full Width	13.75m
Main Engine	2 water-cooled three-phase induction motors
Service Speed	10 knots

Total Floor Area (hotel portion only)
2,639㎡
1st Deck 600m²/2nd Deck 1,136m²/
3rd Deck 903m²

Number of Cabins	19
Number of Passengers	38

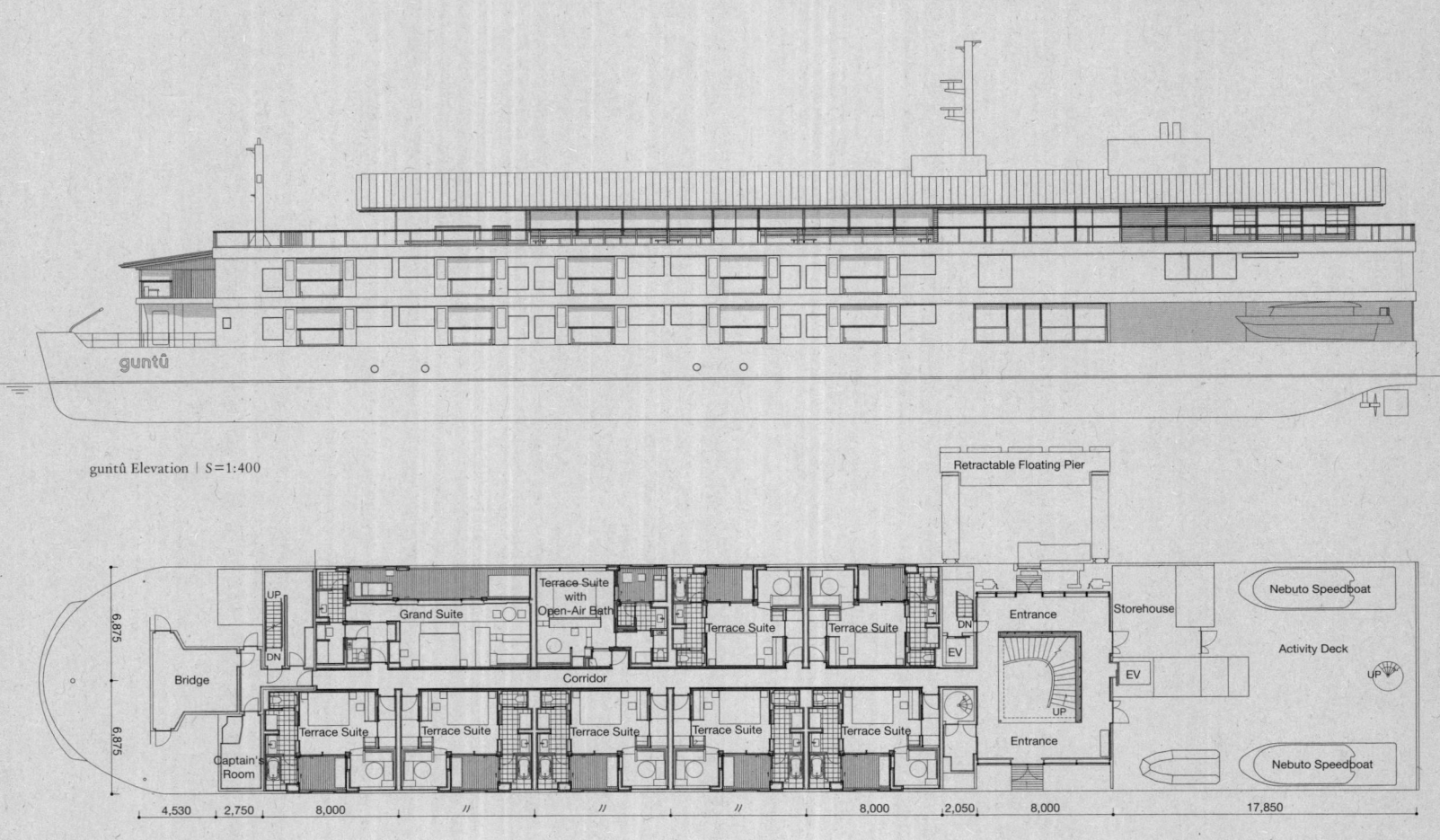

guntû Elevation | S=1:400

1st Deck Plan | S=1:400

ねぶと
Nebuto

総トン数	3.7t	
全長	9.26m	
全幅	2.39m	
主機関	ディーゼル式2機	
運航速度	20ノット	
乗客定員	12名	

Gross Tonnage	3.7t	
Full Length	9.26m	
Full Width	2.39m	
Main Engine	2 diesel engines	
Service Speed	20 knots	
Number of Passengers	12	

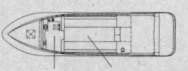

Nebuto Elevation | S=1:400

Cockpit   Passenger Seating   Nebuto Plan | S=1:400

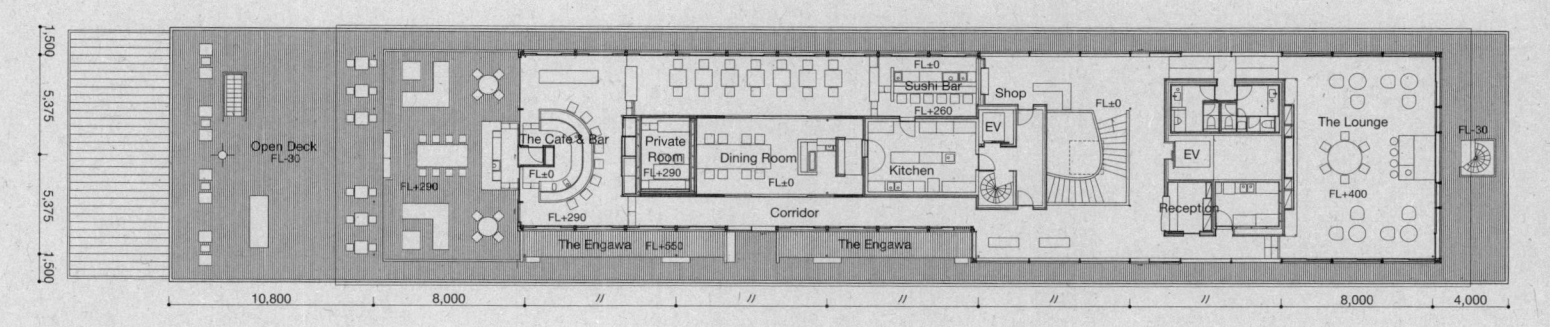

3rd Deck Plan

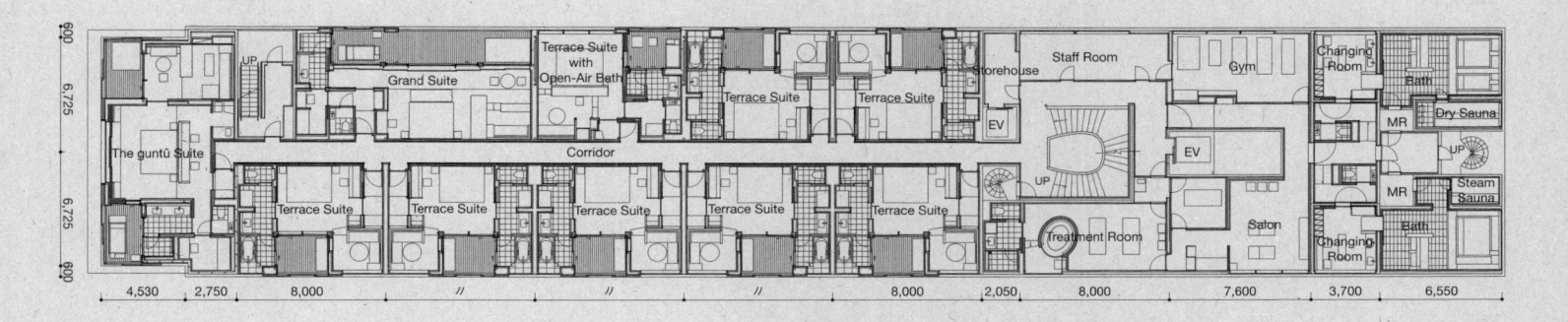

2nd Deck Plan

ガンツウ主要関係者リスト

2017年10月17日就航時

所在地	広島県尾道市浦崎町1364番地6
主要用途	客船
建主	せとうちクルーズ

設計・監理

建築	堀部安嗣建築設計事務所
	[担当]堀部安嗣、伊藤嘉記、小林正義
造船	常石造船
照明	ライティング・プランナーズ・アソシエーツ
	[担当]面出薫、村岡桃子、本多由実
構造協力	多田脩二構造設計事務所 [担当]多田脩二
空調協力	YMO [担当]山田浩幸、関田優子

企画	せとうちクルーズ
和食監修	重よし 佐藤憲三
鮨監修	淡路島 互 坂本互生
和菓子監修	樫舎 喜多誠一郎

施工

造船	常石造船
建築	せとうちホールディングス 建設カンパニー
	[担当]新保智之、松葉秀治
	大和建設
	[担当]小林日出男、大石晋伊智、岡本光正、片岡元広、奥林達也
図	AK Product's [担当]堀川晃
電気	福山電業 [担当]堀尾壮志
衛生	三幸社 [担当]岡村逸輝、清水司
空調	ダイキンMRエンジニアリング
屋根板金	新星商事 [担当]藤井敏尋
木	三洋建材
	[担当]平井利直、桑原俊也、神田祐也、岡田侑大
	越井木材工業 [担当]辻えりか
	岡部材木店 [担当]岡部隆幸
	岡崎製材 [担当]三浦秀尚
左官	営善 [担当]竹内章二
	日丸産業 [担当]笹原康史
塗装	小川塗建 [担当]小川勝
石	WAO [担当]土本昌広
タイル	マイスター・ワコー [担当]和田晃一
	ベイス [担当]横井敦彦
	国代耐火工業 [担当]西脇太亮
金属	トスミック [担当]林伸洋
造作家具	GALLERY-SIGN [担当]溝口至亮
	島田明恵
	アサヒ工芸 [担当]近藤雅彦
	さとう工芸 [担当]佐藤和夫、村田尚光
	泰山堂 [担当]山路宜嗣
造作家具・建具	藤原木工所
	[担当]藤原俊邦、廣川省吾、金高清道
建具	ナニック・ジャパン [担当]武田博美
鋼製建具	大成工業 [担当]開原和政
サッシ	共立機械製作所
ガラス	ADF・アヤベ [担当]綾部系一

内装	マルケン装飾 [担当]岡田展政
サウナ	メトス [担当]又野宏
浴槽	桐山 [担当]桐山明文
照明	トキ・コーポレーション [担当]浅井乙秀
	コイズミ照明 [担当]世良由利子
厨房機器	ホシザキ [担当]藤井辰成
木甲板	巽産業 [担当]青山祐二
	ワイエフエフ
防音	創和設計 [担当]米倉健、鈴江健太
床暖房	GROOVE [担当]丸山貴司、加藤光陽

置き家具	北村武子
	山崎文子
	GALLERY-SIGN [担当]溝口至亮
	コーチ・カズノリ
	hao & mei [担当]傍島浩美
	OZA METALSTUDIO [担当]小沢敦志
	心石工芸 [担当]心石拓男
	イヨベ工芸社 [担当]五百部宗一
	KITAWORKS [担当]木多隆志
	ワイ・エム・ケー長岡 [担当]西脇裕二
	工房さ竹 [担当]佐竹弘章
	マルニ木工 [担当]河村謙一、上杉和敬

サインデザイン／ロゴ・グラフィックデザイン

	平林奈緒美
サイン	香川加奈子
	創工芸 [担当]村上聡
グラフィック	NAOHIRO HAITANI
	小林寛之
空間デザイン	北村武子
	GALLERY-SIGN [担当]溝口至亮
ベッドマット	綿久リネン [担当]中島朗太郎

フィットネスマシン

	テクノジムジャパン [担当]滝川葉子

オーディオ	オーディオテクネ インコーポレイテッド
	［担当］今井清昭
事務機器	サンエイ ［担当］田中修辞
家電	JTB商事 ［担当］二瓶光男
備品デザイン	YUKI PALLIS
	中村有希
	岡崎ちはる
	岩渕友里
	岡崎ちひろ
制服・オリジナルアパレルデザイン製作	
	橋本貴行
	檀上幸伸
	北村武子
絵画	尹熙倉

テンダーボート「ねぶと」
―――

設計	堀部安嗣建築設計事務所
	［担当］堀部安嗣
造船	ツネイシクラフト＆ファシリティーズ

艤装工事中の
ガンツウ。
Guntû during
outfitting.

Photo Credits

鈴木研一
PP.OOI-O40, P.O62, PP.O74-O79,
PP.O82-O90, P.O9I［右上以外］,
PP.O92-O95, P.O96［左］, P.O97,
PP.IOO-IO4, P.IO5［右上以外］, PP.IO6-I2O

堀部安嗣
P.O46, P.O48, P.O5I, P.O6O［下］,
P.O8O, P.O8I［下］, P.O96［右］, P.I25

堀部安嗣建築設計事務所
P.O55［下］, P.O58, P.O6O［上］

伊藤徹也
P.O8I［上］, P.O9I［右上］

テレビマンユニオン
P.O55［上］

富井雄太郎
P.IO5［右上］

［複写］
高橋菜生
PP.O65-O73, O98-O99

［撮影協力］
千光寺山荘｜PP.OO6-OO9, II8-II9
鶴田石材｜PP.II4-II5
吉田邸｜P.IO3［左］

海上保安庁図誌利用第20190005号
PP.O98-O99

P.OOI　静かな内海を航行するガンツウ。
シルバーの船体が風景や光を映す。

PP.OO2-OO5　瀬戸内の夜明け。
島々と海の境界や色が浮き上がってくる。

PP.OO6-OO9　本州と向島の間の尾道水道を行くガンツウ。
尾道水道は狭く、複雑な潮流がある難所。

PP.OIO-OII　向島から尾道の町並みとガンツウを見る。

PP.OI2-OI3　ガンツウから尾道を見る。
陸側手前に建つのは、古い海運倉庫を
ホテルなどが入る複合施設に
リノベーションした「ONOMICHI U2」。

PP.OI4-OI5　切妻屋根が掛かる3階のオープンデッキ。
天井はサワラ、床のデッキはクリ。

PP.OI6-OI7　ガンツウの左舷。
船首の2階に客室ザ ガンツウスイート、
1階に操舵室を配置。
1階と2階は海に面して客室が並び、
アルミ合金製の雨戸を備えている。

PP.OI8-OI9　3階の縁側。尾道水道を望む。
縁側の奥行は1,415mmで、
床は40mm厚のヒノキ。
柱の手が触れる範囲には籐を巻いている。
曲線をもった背板はヒノキ。

P.OI9［右］　3階オープンデッキ。
正面は広島県尾道市の生口島と
愛媛県今治市の大三島をつなぐ
多々羅大橋。

PP.O2O-O2I　航行するガンツウを見下ろす。
屋根はチタン亜鉛合金。

PP.O22-O23　移りゆく風景。その場所ならではの
様々な産業や営み、集落、工場などを見る。

PP.O24-O25　塩飽諸島の中心である本島の集落。
屋根がつくる風景はガンツウの
インスピレーションの源のひとつ。
PP.IOO-IO3 参照。

P.O26　オープンデッキに佇む人。
季節・天候・時間によって様々に変化する
風景と出会う。

P.O27　釣り船とガンツウ。

PP.O28-O29　周囲の光を映し、景色に馴染むガンツウ。

P.O3O　縁側。航行中はわずかな時間でも
光と景色が変化していく。

P.O3I　オープンデッキは3階の四周にある。
手摺りは手触りの良いヒノキ。

PP.O32-O33　ガンツウの右舷。
船尾にテンダーボート「ねぶと」
2艘を積載し、その乗降のための
開閉式浮き桟橋を備えている。
写真は浮き桟橋が降りた状態。

PP.O34-O35　瀬戸内独特の夕景。
基本的に夕方からは錨泊し、
日暮れの時間を楽しむことができる。

PP.O36-O37　オープンデッキ夕景と夜明け。

PP.O38-O39　オープンデッキ夜景。
3階の切妻屋根の下では、
軒や天井と床レベルの違いにより
それぞれの場所で異なる体感、
視界を得られる。

P.O4O　ガンツウ船尾と多々羅大橋。

Photo Credits

———

Ken'ichi Suzuki
PP.001–040, P.062, PP.074–079,
PP.082–090, P.091 [except upper right],
PP.092–095, P.096 [left], P.097,
PP.100–104, P.105 [except upper right],
PP.106–120

Yasushi Horibe
P.046, P.048, P.051, P.060 [bellow],
P.080, P.081 [bellow], P.096 [right], P.125

———

Yasushi Horibe Architect & Associates.
P.055 [bellow], P.058, P.060 [above]

———

Tetsuya Ito
P.081 [above], P.091 [upper right]

———

TV MAN UNION, INC.
P.055 [above]

———

Yutaro Tomii
P.105 [upper right]

[Image reproduction]
Nao Takahashi
PP.065–073, 098–099

[Photography cooperation]
SENKOJI SANSO | PP.006–009, 118–119
TSURUTA STONE. Co., Ltd. | PP.114–115
Yoshida House | P.103 [left]

P.001
Guntû navigating the placid Inland Sea. The silver hull reflects the scenery and light.

PP.002–005
Dusk in Setouchi. The outlines and colors of the islands and sea stand out.

PP.006–009
Guntû passes through the Onomichi Channel between Honshu and Mukojima. The Onomichi Channel is a challenging spot due to its narrowness and complex tides.

PP.010–011
Guntû and the cityscape of Onomichi seen from Mukojima.

PP.012–013
Onomichi seen from guntû. Standing along the shore is Onomichi U2, an old maritime shipping warehouse that was renovated into a hotel and mixed-use facility.

PP.014–015
The third-floor open deck covered by the gabled roof. The ceiling is finished with sawara cypress and the deck with chestnut.

PP.016–017
Guntû's port side. The bow of the ship features the guntû Suite on the second deck and the bridge on the first deck. Cabins face the sea on both the first and second decks and are equipped with aluminum alloy storm shutters.

PP.018–019
The Engawa on the third deck, looking towards the Onomichi Channel. The Engawa has a depth of 1,415mm and 40mm-thick hinoki floors. The columns are wrapped with cane at hand level. The backrests are made from curved hinoki.

P.019 [right]
The third-floor open deck. Ahead is the Tatara Bridge connecting Ikuchijima in Onomichi, Hiroshima and Omishima in Imabari, Ehime.

PP.020–021
Looking down on guntû while underway. The roof is made of titanium-zinc alloy.

PP.022–023
The shifting scenery includes views of a wide variety of local industry, settlements, and factories.

PP.024–025
A village on Honjima at the center of the Shiwaku Islands. Landscapes formed by roofs are one source of inspiration for guntû. See pp.100-103.

P.026
Standing on the open deck brings encounters with scenery that changes with the season, weather and hour of day.

P.027
Fishing boat and guntû.

PP.028–029
Guntû reflects the surrounding light and blends into the scenery.

P.030
The Engawa. The light and scenery can change in a few moments when underway.

P.031
Open decks line all four sides of the third deck. Railings are made from smooth-to-touch hinoki.

PP.032–033
Guntû's starboard side. The ship's stern holds two "Nebuto" speedboats and is also outfitted with a retractable floating pier for boarding and disembarking from the speedboats. The pier is extended in the photograph.

PP.034–035
A distinctive Setouchi sunset. The ship usually anchors in the evening, allowing enjoyment of the dusk hours.

PP.036–037
Sunset and daybreak on the open deck.

PP.038–039
Night on the open deck. Beneath the gabled roof on the third deck, the bodily sensation and field of vision change at each point depending on the different levels of the eaves, ceiling, and floor.

P.040
Guntû's stern and the Tatara Bridge.

ガンツウ | guntû

2019年8月19日　初版第1刷　　1st edition: 19 August 2019

著者　　　　　　　　　　Author
堀部安嗣　　　　　　　　Yasushi Horibe

写真　　　　　　　　　　Photographer
鈴木研一　　　　　　　　Ken'ichi Suzuki

協力　　　　　　　　　　Cooperation
せとうちクルーズ　　　　SETOUCHI CRUISE, INC.
堀部安嗣建築設計事務所　Yasushi Horibe Architect & Associates.

翻訳　　　　　　　　　　Translation
サム・ホールデン　　　　Sam Holden

デザイン　　　　　　　　Design
小池俊起　　　　　　　　Toshiki Koike

印刷・製本　　　　　　　Printing and binding
図書印刷　　　　　　　　TOSHO Printing Co., Ltd.

編集・発行　　　　　　　Editing and publishing
富井雄太郎　　　　　　　Yutaro Tomii

発行所　　　　　　　　　Publisher
millegraph　　　　　　　 millegraph

millegraph
—
tel & fax　+81-(0)3-5848-9183
mail　　　info@millegraph.com
http://www.millegraph.com

ISBN 978-4-910032-00-9 C-0052
Printed in Japan.